The Asian Hornet in Europe

An Unforeseen Invasion

By

Julien Fortanelle

Table of contents :

INTRODUCTION

For some years now, Europe and the United Kingdom have been facing a formidable invader: the Asian hornet (Vespa velutina). Originating from Asia, this predatory insect was accidentally introduced to Europe, triggering a serious threat to biodiversity and the local ecosystem. Its alarming presence and aggressive behavior make it a topic that captures headlines in many countries.

In this first part of our book, we aim to thoroughly explore the issue of the Asian hornet, highlighting its characteristics, its impact on nature and humans, and the measures taken to manage this invasion. We will delve into the origins of this invasive insect, its introduction to Europe, and the reasons that facilitated its rapid spread in different regions of the continent.

Firstly, we will discover the peculiarities of the Asian hornet, distinguishing it from its native counterparts. Its unique appearance and social behavior intrigue scientists, who are trying to better understand how this invasive species has thrived in a new environment.

We will then analyze the devastating consequences of the Asian hornet on local biodiversity. Many ecosystems are disrupted by its presence, causing a decline in pollinator insect populations and disrupting natural food chains. Indigenous species, unable to cope with this new threat, suffer significant harm, jeopardizing the delicate balance of fauna and flora.

Beyond its effects on nature, the Asian hornet also poses a risk to the health and safety of human populations. Its stings, more numerous and potentially more dangerous than those of other hornet species, can cause severe allergic reactions. Beekeepers, who rely on bees for crop pollination, are particularly affected, while farmers see their yields decrease due to the predation of the Asian hornet on harmful insects.

Faced with this growing threat, many European countries have undertaken management and control efforts to contain the spread of the Asian hornet. Nest eradication techniques have been implemented, and surveillance programs have been established to identify colonies before they become uncontrollable.

In this work, we will explore these management measures in detail, while providing practical advice on how to deal with an Asian hornet nest and potential stings. It is essential that each individual is aware of this issue and actively participates in environmental preservation by adopting appropriate practices.

In conclusion, this book aims to raise awareness among readers about the challenges related to the Asian hornet in Europe and the United Kingdom. By understanding the risks and consequences of its presence, we can better grasp this invasion and work together to protect our environment and health.

Chapter 1: The Asian Hornet in a Few Words

The Asian hornet (Vespa velutina) is an insect species belonging to the Vespidae family. Native to Asia, it is also known as the yellow-legged hornet due to its characteristic yellow legs, which distinguish it from indigenous European hornet species.

Physical Description and Distinctive Features

The Asian hornet is easily recognizable by its imposing size. Workers generally measure between 2.5 and 3 centimeters in length, while queens can reach up to 3.5 centimeters. Its coloration is dark, with a dominant black hue and orange or yellow bands on the abdomen. The previously mentioned yellow legs, as well as its orange face, are essential distinctive traits for identification.

Furthermore, the Asian hornet has brown wings and a wide, flat head with large antennae. Its thorax is black, scattered with small yellow spots on the sides. In contrast to European hornets, it sports an orange band on the last segment of its abdomen, facilitating its identification.

Geographical Origin and Introduction to Europe

The Asian hornet originates from Southeast Asia, mainly from southern China and northern India. It was accidentally introduced to Europe, likely in a shipment of pottery or materials from Asia, during the years 2004-2005. This unintentional introduction in France marked the beginning of the rapid spread of the Asian hornet across the European continent.

Comparison with Indigenous Hornet Species

In Europe, the Asian hornet significantly differs from indigenous hornet species, such as the European hornet (Vespa crabro) and the German hornet (Vespa germanica). Besides its slightly larger size, it exhibits distinctive features, notably its yellow legs and orange bands on the abdomen.

In terms of behavior, the Asian hornet also stands out for its predatory diet, with a particular focus on hunting insects, primarily bees and wasps, making it a formidable predator for bee colonies.

This first part of the book has allowed us to explore the specifics of the Asian hornet, its Asian origin, and its accidental introduction to the United Kingdom, as well as the differences that distinguish it from indigenous hornet species. In the next chapter, we will address the devastating ecological impact of this invasive species on biodiversity and the local ecosystem.

Chapter 2: The Ecological Impact of the Asian Hornet

The Asian hornet (Vespa velutina) has proven to be an exceptionally efficient invasive predator, causing significant disruptions within local ecosystems in the United Kingdom. Its accidental introduction has had dramatic consequences on biodiversity, leading to undesirable changes in the ecological balances of the affected areas.

Influence on Local Ecosystems and Biodiversity

The presence of the Asian hornet can disturb the delicate balance of local ecosystems. Due to its predatory diet, it exerts considerable pressure on insect populations, particularly bees, wasps, and other pollinating insects. Domestic bee colonies, crucial for plant pollination and agricultural yields, are particularly vulnerable to the attack of the Asian hornet.

Moreover, as a formidable predator, the Asian hornet can destabilize populations of other insects, impacting the food chain and interactions between species. This can have repercussions on local fauna and flora, affecting insectivores and animals dependent on resources provided by insects.

Consequences for Pollinating Insect Populations

Bees play a crucial role in plant pollination, contributing to the reproduction of many plant species, including food crops. The

Asian hornet is a formidable predator for these essential pollinators. It locates bees in flight, captures worker bees at hive entrances, and then kills them to feed on their abdomens, leaving behind a decimated hive.

The predation of Asian hornets on bees can lead to a significant decrease in bee populations, compromising the pollination of cultivated and wild plants. This situation jeopardizes food production, agricultural economies, and the diversity of local plant species.

Impacts on Local Fauna and Flora

Beyond pollinating insects, the Asian hornet can also impact other forms of life in local ecosystems. Its predation on insects can destabilize populations of other predators and disrupt food chains, resulting in ecological imbalances. Insect scarcity can affect insect-eating birds and other animals dependent on this food resource.

Additionally, declines in certain plant species may be observed due to reduced pollination. This situation can have consequences on the composition of local plant communities and the dynamics of ecosystems.

This chapter has highlighted the devastating repercussions of the Asian hornet on local ecosystems in Europe. From its influence on biodiversity to its impact on pollinating insect populations and the consequences for local fauna and flora, the Asian hornet poses a serious threat to the stability of ecosystems. In the next chapter, we will explore the dangers that this invader poses to humans, as well as prevention and safety measures to limit the associated risks of its presence.

Chapter 3: Consequences for Humans

The Asian hornet (Vespa velutina) poses significant risks to human health and safety, especially due to its potentially dangerous stings. Its expansion in the United Kingdom has raised concerns about allergic reactions and impacts on human activities, including beekeeping and agriculture.

Risks Associated with Asian Hornet Stings: Allergies and Reactions

Asian hornet stings can cause local reactions such as redness, itching, and swelling at the sting site. However, for some individuals, these stings can lead to severe allergic reactions, ranging from generalized urticaria to potentially life-threatening anaphylactic shock. Symptoms of a severe allergic reaction may include difficulty breathing, swelling of the face and throat, a general feeling of discomfort, and a drop in blood pressure.

In regions where the Asian hornet is well-established, medical services must be prepared to handle emergency cases related to stings from this insect. It is crucial to raise public awareness about the risks associated with Asian hornet stings and provide information on the symptoms of an allergic reaction for prompt and appropriate management.

Threats to Beekeepers and Farmers

Beekeepers are particularly vulnerable to the Asian hornet due to its repeated attacks on bee colonies. The predator approaches hives to capture worker bees at the entrance, causing significant stress to the colonies. When Asian hornets are abundant, they can weaken a hive or even destroy it entirely, compromising honey and pollen production.

Farmers also experience indirect consequences of the presence of the Asian hornet, as it is a formidable predator of pests harmful to crops. The decrease in populations of pollinating insects, especially bees, can affect crop pollination, leading to reduced yields and a loss of crop quality.

Preventive and Safety Measures to Minimize Risks

To limit risks associated with the Asian hornet, preventive and safety measures must be adopted. People living in areas where the Asian hornet is present are advised to exercise caution and avoid disturbing nests. In the event of discovering an Asian hornet nest, it is essential to report its presence to the relevant authorities so that a specialized team can safely intervene for its elimination.

Beekeepers can take measures to protect their hives by installing specific traps for

Asian hornets and reinforcing hive entrances to limit predator access. Special attention should also be paid to signs of increased Asian hornet activity around hives to react quickly in the event of a threat.

For individuals with allergies, it is essential to consult a healthcare professional and have an emergency kit on hand to treat potential allergic reactions following a sting.

This chapter has highlighted the serious consequences that the Asian hornet can have for humans, particularly regarding potentially allergic stings and threats to beekeepers and farmers. To minimize risks, preventive and safety measures must be adopted, and the public must be informed about how to respond to the presence of this predatory insect. In the next chapter, we will explore the spread of the Asian hornet in Europe and the factors promoting its rapid dissemination.

Chapter 4: The Spread of the Asian Hornet in Europe

Since its accidental introduction in France, the Asian hornet (Vespa velutina) has rapidly expanded across the country and the European continent, raising concerns about its dissemination and adaptation to new environments. This chapter examines the mapping of its expansion, the factors that have facilitated its spread, and the surveillance and monitoring initiatives implemented to better understand and manage this invasion.

Mapping its Expansion in Europe and the United Kingdom and Migration Routes

The progress of the Asian hornet in the United Kingdom has been swift since its introduction. Studies and reports show that the species has extended its distribution to various regions, including England, Scotland, Wales, and Northern Ireland. The mapping of this expansion allows for a visual representation of the species' geographic evolution over the years.

The migration routes of the Asian hornet are diverse and can be facilitated by human movement, the transport of goods, and climatic conditions conducive to its dissemination. Founding queens can travel long distances, establishing new colonies and satellite colonies at each stage of their journey.

Factors Favoring its Dissemination and Adaptation to New Environments

The Asian hornet has demonstrated a remarkable ability to adapt to diverse habitats, ranging from rural to urban areas. Its generalist diet allows it to find varied food resources, facilitating its adaptation to different environments.

Moreover, the absence of natural predators in the United Kingdom has favored the growth of Asian hornet populations. Unlike indigenous species, it faces no major competitors for food resources, enabling it to thrive rapidly.

Furthermore, the high reproductive capacity of Asian hornets, combined with their aggression towards other species, has allowed them to gain an advantage over local species.

Surveillance and Monitoring Initiatives

Confronted with the growing threat of the Asian hornet, numerous surveillance and monitoring initiatives have been implemented to better understand the dynamics of its populations and anticipate its spread. Collaborative monitoring networks have been established, where citizens and scientists work together to report observations of Asian hornets and their nests.

These monitoring initiatives aim to identify the most affected areas, detect colonies early, and better understand the migratory behaviors of the species. This information is crucial for developing effective management and control strategies.

In conclusion, this chapter has allowed us to understand the rapid spread of the Asian hornet in Europe and the United Kingdom and the factors that have facilitated its adaptation to new environments. Mapping its expansion and surveillance initiatives are essential for better understanding this invasion and implementing appropriate measures to limit its impact on the environment and human activities. In the next chapter, we will examine methods for managing Asian hornet nests and colonies, as well as preventive actions to minimize the risks associated with their presence.

Chapter 5: Management of Asian Hornet Nests and Colonies

The management of Asian hornet nests and colonies is a major challenge to mitigate the impact of this invasive species on the environment and human safety. In this chapter, we will explore methods to identify and report Asian hornet nests, as well as appropriate and secure elimination techniques. Practical advice will also be provided to prevent stings and ensure effective protection.

Identification and Reporting of Asian Hornet Nests

The first step in managing Asian hornets is to know how to identify their nests. These nests are typically located at height, in trees, bushes, attics, building cavities, or under roofs. The nests have a spherical or oval shape, constructed from paper mache using wood fibers and saliva.

It is essential to raise public awareness about recognizing Asian hornet nests and differentiating them from nests of other indigenous wasps and hornets. Upon discovering a suspicious nest, it is recommended to report its presence to the relevant authorities, such as local fire and rescue services, pest control associations, or environmental services.

Appropriate and Secure Nest Elimination Techniques

The destruction of Asian hornet nests must be carried out with caution and by trained professionals, as it can be dangerous for inexperienced individuals. Specialists use specific methods to eliminate nests based on their location and accessibility. They wear personal protective equipment to guard against potential stings.

Several nest elimination techniques exist, including the use of sprayers, insecticidal powder, or selective traps. The goal is to eliminate the queen and the colony to prevent their multiplication. It is crucial not to disturb the nests or attempt self-elimination, as this could trigger defensive attacks from Asian hornets, jeopardizing the safety of individuals.

Prevention and Protection against Stings: Practical Tips

To minimize the risks of Asian hornet stings, preventive measures can be adopted. It is recommended to avoid approaching nests or areas where Asian hornets are active. During outdoor activities such as gardening, hiking, or picnics, wearing covering clothing and light colors is advisable, as this can help deter Asian hornets from feeling threatened.

Insect repellents can also be useful in discouraging Asian hornets from approaching. In the presence of nests in the vicinity, it is essential to inform neighbors, local communities, and relevant authorities so that appropriate management measures can be implemented.

In conclusion, this chapter has explored the management of Asian hornet nests and colonies, focusing on the identification and reporting of nests, appropriate and secure elimination techniques, as well as practical advice to prevent stings and ensure individual protection. In the next chapter, we will address the consequences of the Asian hornet invasion on ecosystems and local populations of pollinating insects, along with preventive actions to limit its expansion.

CONCLUSION

In this book, we have delved into the profound impact of the Asian hornet (Vespa velutina) in Europe and United Kingdom, an invasive predator that has made headlines in recent years. We have covered several key aspects of this issue, highlighting environmental challenges, risks to humans, and the necessary management and prevention measures to address this invasion.

We began with a brief introduction to the Asian hornet, emphasizing its distinctive characteristics and geographical origin. Its accidental introduction to Europe paved the way for rapid expansion, disrupting local ecosystems and endangering biodiversity.

The ecological impact of the Asian hornet was examined in detail, revealing devastating consequences for populations of pollinating insects, food chains, as well as local fauna and flora. Beekeepers and farmers were also identified as vulnerable groups, facing repeated attacks from this predatory insect.

We then explored the spread of the Asian hornet in Europe and United Kingdom, highlighting migration routes, factors favoring its dissemination, and adaptation to new environments. Population surveillance and monitoring initiatives were emphasized as essential tools to better understand this invasion and develop adequate management strategies.

Addressing the management of Asian hornet nests and colonies, we underscored the importance of nest identification and reporting, as well as the use of appropriate and secure techniques for their elimination. Prevention and protection against stings were also discussed to limit the risks associated with the presence of Asian hornets.

Faced with this growing threat, it is essential that we collectively take effective action against the Asian hornet in Europe and United Kingdom. We call for action, raising awareness among the public, authorities, beekeepers, farmers, and all stakeholders about the importance of taking measures to limit its expansion.

Collaboration among all these stakeholders is paramount in combating this invasion. Together, we can strengthen surveillance and monitoring initiatives, coordinate consistent management strategies, and implement appropriate prevention programs. Only concerted and coordinated action will effectively protect the environment, local species, and human health.

In conclusion, we must demonstrate determination and commitment to meet the challenge of the Asian hornet. By adopting a proactive approach, we can preserve our biodiversity, ecosystems, and quality of life in the face of this persistent threat. Together, we can take measures to limit the expansion of the Asian hornet and protect our natural environments and communities.